Stevia

NATURE'S SWEETENER

Rita Elkins, M.H.

WOODLAND PUBLISHING
Pleasant Grove, UT

© 1997
Woodland Publishing, Inc.
P.O. Box 160
Pleasant Grove, UT
84062

The information in this book is for educational purposes only and is not recommended as a means of diagnosing or treating an illness. All matters concerning physical and mental health should be supervised by a health practitioner knowledgeable in treating that particular condition. Neither the publisher nor author directly or indirectly dispense medical advice, nor do they prescribe any remedies or assume any responsibility for those who choose to treat themselves.

Contents

Introduction	5
A Brief History	6
Artificial Sweeteners: Cause For Worry	10
The FDA and Non-Caloric Sweeteners	12
Sugar, Sugar Everywhere	15
Stevia: The Ideal Sweetener?	19
Forms of Stevia	24
How To Use Stevia	25
Safety	26
Primary Applications of Stevia	27
Conclusion	27
Bibliography	29

Introduction

While aspartame and saccharine continue to dominate the non-caloric sweetener scene, a remarkable herb called stevia remains relatively obscure. Why would a substance that is much sweeter than sugar, can be used in baking (unlike aspartame), is diabetic-safe and calorie-free remain unknown and unused? Unfortunately, the FDA has managed to unfairly keep stevia out of the American market due to a long history of unwarranted regulation. Recently, however, due to the passage of specific legislation, limited purchase of stevia products is now available.

Anyone who suffers from blood sugar disorders or who needs to limit their caloric intake should know about the remarkable properties of stevia. Stevia offers an ideal alternative to other sugars or sugar substitutes. Moreover, the herb has numerous therapeutic properties and has proven its safety and efficacy for hundreds of years.

In spite of FDA efforts to ban this herbal sweetener, stevia's comeback has begun amidst a glut of approved artificial, pharmaceutical products that pose significant health risks. The story of stevia illustrates the struggle which many natural products have experienced in gaining the FDA stamp of approval. Patents, politics and profits are all involved in determining the selection of products we are allowed to purchase. The history of stevia's use in this country epitomizes the sad fact that effective natural supplements are often suppressed, while much riskier artificial chemicals are praised and aggressively marketed.

STEVIA (*Stevia rebaudiana*)

SYNONYMS: sweet herb, honey leaf
PARTS USED: leaves

Description

Stevia is a small perennial shrub with green leaves that belongs to the aster (Asteraceae) or chrysanthemum family of plants. They grow primarily in the Amambay mountain range of Paraguay but over 200 various species of stevia have been identified around the globe. *Stevia rebaudiana* is the only species at present which possesses an inordinate ability to sweeten. Its common form is known as stevioside, a fine white powder extracted from the leaves of the plant.

Phytochemistry

STEVIOSIDE/REBAUDIDOSIDE COMPOUND DUO: The leaves of the stevia shrub contain specific glycosides which produce a sweet taste but have no caloric value. Stevioside is the primary glycoside involved in this effect. Dulcoside and rebaudioside are also major glycosides contained in the herb. Glycosides are organic compounds which contain a sugar component (glycone) and a non-sugar component (aglycone). The glycone constituent may be comprised of rhamnose, fructose, glucose, xylose, arabinose etc. The other portion may be any kind of chemical compound such as a sterol, tannin, carotenoid, etc.

Stevia leaves also contain protein, fibers, carbohydrates, phosphorus, iron, calcium, potassium, sodium, magnesium, rutin (flavonoid), iron, zinc, vitamin C and vitamin A. Human physiology cannot metabolize the sweet glycosides contained in stevia leaves, therefore they are eliminated from the body with no caloric absorption. Stevia, unlike aspartame, can be used in baking because its sweet glycosides do not break down when heated.

Definition

Stevia is an herb with incredible sweetening power. Its ability to sweeten is rated between 70 to 400 times that of white sugar. Typically, it has a mild licorice-like taste and is completely natural in its biochemical profile. What makes stevia so intriguing is that unlike other natural sweetening agents, its is completely calorie-free, never initiates a rise in blood sugar, and does not provide "food" for microorganisms like bacterias and yeasts.

Stevia may well be the most remarkable sweetener in the world and yet its recognition in this country remains relatively low. Consider the extraordinary attributes of the stevia plant and its extracts:

- It is diabetic-safe.
- It is calorie-free.
- It is 50 to 400 times sweeter than white sugar.
- It does not adversely effect blood sugar levels.
- It is non-toxic.
- It inhibits the formation of cavities and plaque.
- It contains no artificial ingredients.
- It can be used in baking and cooking.

A Brief History

Stevia is a plant indigenous to mountainous regions of Brazil and Paraguay. For centuries, this herbal sweetener has been used by native cultures to counteract the bitter taste of various plant-based medicines and beverages. The Guarani Indians of Paraguay have used this potent sweetener in their green tea for generations. The name they designated for stevia leaves was "sweet herb." In addition, these native peoples have historically used stevia as a digestive aid and a topical dressing for wounds and other skin disorders.

In the sixteenth century, Europeans became aware of the herbal sweetener through the Spanish Conquistadors. In the late 1880s, Moises S. Bertoni, director of the College of Agriculture in Asunción, Paraguay, became extremely intrigued by the stevia plant. Its reputation was that it was so sweet that even just a small leaf part could sweeten an entire container of mate tea. Bertoni wanted to find out if this was true.

After several years of studying the plant, he wrote about it in a local botanical publication. In 1905, Bertoni published an important article about the incredible sweetening power of the stevia plant, which he considered superior to sugar and extremely marketable. Other articles written by Bertoni note that stevia is unquestionably superior to saccharine because it is nontoxic and has significant therapeutic benefits. It sweetens with unprecedented potency and can be used in its natural state.

The first stevia crop was harvested in 1908 and subsequently, stevia plantations sprang up in South America. In 1921, the American Trade Commissioner to Paraguay, George S. Brady, wrote that although the herb is an extraordinary sweetener with remarkable properties, little had been done to commercially cultivate the plant. He suggested that stevia may be an ideal sugar product for diabetics and strongly advised that American companies pursue its importation.

During the decade of the 1970s, the Japanese developed a new method which could better refine the glycosides contained in the stevia leaf. The result was a compound called stevioside which is from 200 to 300 times sweeter than white sugar. The Japanese approach artificial sweeteners with great caution and they believe stevioside to be safer and more effective than other non-nutritive, chemical products. Stevioside is considered superior in its ability to sweeten; however, it does not exhibit some of the other therapeutic actions found in whole stevia leaves.

Stevia enjoyed substantial popularity during the 1980s as a natural sweetener and was found in a variety of consumer products. In 1986, however, the FDA abruptly seized stevia inventories and in 1991 claimed it was not suitable as a food additive. Advocates for stevia claim this happened because the herb is a natural, powerful, inexpensive and non-patentable sweetener, and therefore poses a threat to pharmaceutical sweeteners and sugar-alcohol sweeteners like mannitol, sorbitol and xylitol. At this writing, stevia has received approval by the FDA to be sold only as a dietary supplement, not as a sweetening agent.

Currently, stevia is commercially grown in Paraguay, Brazil, Uruguay, Central America, Israel, China, Thailand, and the United States. It is considered an important natural sweetener in both Japan and Korea, and has been safely used in these countries for decades. Extracts of stevia and related products make up a considerable portion of the Japanese market for natural sweetening agents. They use stevia in sweet sauces, pickles, beverages, etc., making Japan one of the largest single consumers of stevia in the world.

Today, because the demand for stevia is escalating, several Paraguayan organizations are looking to expand the commercial cultivation of the plant. Currently, Canadian researchers and chemists are working to provide even better stevia supplements and may even end up teeming with governmental agencies to raise stevia crops as economic replacements for tobacco leaves (Bonvie, 64). Stevia has not been officially approved by Canadian agencies, but it is still available for purchase in tea form.

ARTIFICIAL SWEETENERS: CAUSE FOR WORRY

Among some of the most troubling food additives that we routinely ingest are artificial sweeteners, also referred to as non-nutritive sweeteners. Having received the FDA stamp of approval, they are liberally ingested with little thought to what their actual health risks may be. Andrew Weil, M.D., in his book *Natural Health Natural Medicine*, writes:

More worrisome than preservatives are artificial sweeteners. Saccharin, a known carcinogen, should be avoided. Cyclamates, banned some years ago for suspected carcinogenicity, are not being reconsidered for use in food. They taste better than saccharin but cause diarrhea in some people. Avoid them too. Recently, aspartame (NutraSweet) has become enormously popular. The manufacturer portrays it as a gift from nature, but, although the two component amino acids occur in nature, aspartame itself does not. Like all artificial sweeteners, aspartame has a peculiar taste. Because I have seen a number of patients, mostly women, who report headaches from this substance, I don't regard it as free from toxicity. Women also find that aspartame aggravates PMS (premenstrual syndrome). I think you are better off using moderate amounts of sugar than consuming any artificial sweeteners on a regular basis. A natural sweetener that may cause some people problems is sorbitol, originally derived from the berries of the mountain ash tree. Sorbitol tastes sweet but is not easily absorbed form the gastrointestinal tract and is not easily metabolized. It is a common ingredient of sugarless chewing gums and candies. If you eat a lot of it, you will probably get diarrhea. People with irritable bowel syndrome or ulcerative colitis should avoid sorbitol.

Ann Louise Gittleman, in her book, *Super Nutrition for Women*, writes:

In 1977, a Canadian study indicated that when pregnant rats were fed large doses of saccharin, their male offspring developed bladder cancer. As a result, the Canadians banned saccharin and the U.S. Congress ordered warning labels on all saccharin products like Sweet 'N Low. The national Academy of Sciences in 1978 evaluated the evidence and concluded that saccharin was primarily a promoter of other cancer-causing agents, a cocarcinogen. In the meantime, G.D. Searle developed aspartame, a combination of two amino acids and methanol (wood alcohol) . . . Few long-term studies of the effects of aspartame have been done. However, reports to the Food and Drug Administration and the Centers for Disease Control indicate that, as more people consume the substitute in large quantities, health may be affected. In some circumstances, individuals may be getting high levels of methanol; for example, it is estimated that on a hot day after exercise, an individual drinking three 12-ounce cans of diet cola could easily consume as much as eight times the Environmental Protection Agency's recommended limit for methanol consumption. The most common complaints are dizziness, disorientation, tunnel vision, ear buzzing, loss of equilibrium, numbing of hands and feet, inflammation of the pancreas, high blood pressure, eye hemorrhages and seizures. Artificial sweeteners can stimulate hunger or cause additive allergies, just as sugar does. In other words, we get the disadvantages of sugar, along with the proven or suspected disadvantages of artificial sweeteners.

While thousands of Americans continue to consume aspartame in unprecedented amounts, controversy surrounding its safety lingers. Dr. Richard Wurtman of the Massachusetts Institute of Technology (MIT) has reported that abnormal concentrations of neurotransmitters developed when he fed laboratory animals large doses of aspartame. He believes that the phenylalanine content of the sweetener actually manipulates and alters certain brain chemicals which could initiate behavioral changes and even seizures. He also purports that while small quantities of aspartame may be safe, the cumulative effects of the compound—particularly if consumed with high carbohydrate, low protein snacks—could be serious (Wurtman I, 799-801, Wurtman II, 429-430, Wurtman III, 1060).

In spite of serious concerns, saccharine and aspartame packets sit in restaurant sugar bowls all over our country, while in Japan, natural stevia powder enjoys popularity as one of the best and safest non-caloric sweeteners available.

THE FDA AND NON-CALORIC SWEETENERS

While white sugar, turbinado, fructose, honey and corn syrup all qualify as natural sweeteners, none of these are calorie-free nor can they be used by people who suffer from blood sugar disorders. They can encourage weight gain, tooth decay, raise blood sugar quickly, and can also predispose certain individuals to yeast infections. These sugars can also contribute to indigestion, bowel disorders and, possibly, hyperactivity or ADD in children.

Pharmaceutical sweeteners like aspartame and saccharin qualify as calorie-free but come with significant limitations and health risks. Saccharin has been labeled with a warning that it has caused the development of cancer in laboratory animals but is still available for purchase. In 1970, cyclamates, another class of artificial sweeteners, were banned because of the strong possibility that they are, in fact, carcinogenic.

Aspartame has been marketed as a safe substance for the general public, except for those few individuals who suffer from PKU (phenylketonuria), a relatively rare disorder. Most consumers assume that aspartame is a perfectly benign compound and use it liberally. It is, in fact, comprised of phenylalanine, aspartic acid, and methanol (wood alcohol). As previously mentioned, various side effects have been associated with the ingestion of aspartame and include migraines, memory loss, slurred speech, dizziness, stomach pain, and even seizures.

In addition, because aspartame contains chemicals which affect brain cell function, significant questions have been raised concerning its link to increased incidence of brain tumors (Olney). Acesulfame K, another artificial sweetener on the market, has also been linked to cancer by the Center for Science in the Public Interest. Despite the protest of various organizations and health professionals, these pharmaceutical sweeteners have been approved by the FDA and are recognized as safe.

THE FDA AND STEVIA

While stevia in no way qualifies as an "artificial sweetener," it has been subject to rigorous inquiry and unprecedented restraints. In 1986, FDA officials began to investigate herb companies selling stevia and suddenly banned its sale, calling it "an unapproved food additive." Then in 1991, the FDA unexpectedly announced that all importation of stevia leaves and products must cease, with the exception of certain liquid extracts which are designed for skin care only. They also issued formal warnings to companies and claimed that the herb was illegal. The FDA was unusually aggressive in its goal to eliminate stevia from American markets, utilizing search and seizure tactics, embargoes and import bans. Speculation as to why the FDA intervened in stevia commerce points to the politics of influential sugar marketers and the artificial-sweetener industry.

During the same year, the American Herbal Products Association (AHPA) began their defense of the herb with the goal of convincing the FDA that stevia is completely safe. They gathered documented literature and research on both stevia and other non-caloric sweeteners. The overwhelming consensus was that stevia is indeed safe, and the AHPA petitioned the FDA to exempt stevia from food additive regulations.

Food Additive vs. Dietary Supplement

FDA regulations of stevia were based on its designation as a food additive. The claim was that scientific study on stevia as a food additive was inadequate. Ironically, extensive Japanese testing of stevia was disregarde—regardless of the fact that this body of documented evidence more than sufficiently supported its safe use. Many experts who have studied stevia and its FDA requirements have commented that the FDA wants far more proof that stevia is safe than they would demand from chemical additives like aspartame.

Stevia advocates point out that stevia not a food additive, but rather, a food. Apparently, foods that have traditionally been consumed do not require laborious and expensive testing for safety under FDA regulations. The fact that so many toxicology studies have been conducted in Japan, coupled with the herb's long history of safe consumption, makes a strong case for stevia being accepted by the FDA as a safe dietary substance. Still, it was denied the official GRAS (generally recognized as safe) status and designated a food additive by the FDA.

The FDA Reverses Its Position

As a result of the Health Freedom Act passed in September of 1995, stevia leaves, stevia extract, and stevioside can be imported to the United States. However, ingredient labels of products that contain stevia must qualify as dietary supplements. Stevia had been redesignated as a dietary supplement by the FDA and consequently can be legally sold in the United States solely as a supplement. Its addition to teas or other packaged foods is still banned. Moreover, stevia cannot, under any circumstances, be marketed as a sweetener or flavor enhancer.

SUGAR, SUGAR EVERYWHERE

Ralph Nader once said, "If God meant us to eat sugar, he wouldn't have invented dentists." The average American eats over 125 pounds of white sugar every year. It has been estimated that sugar makes up 25 percent of our daily caloric intake, with soda pop supplying the majority of our sugar ingestion. Desserts and sugar-laden snacks continually tempt us, resulting in an escalated taste for sweets.

The amount of sugar we consume has a profound effect on both our physical and mental well-being. Sugar is a powerful substance which can have drug-like effects and is considered addictive by some nutritional experts. William Duffy, the author of *Sugar Blues,* states,"The difference between sugar addiction and narcotic addition is largely one of degree."

In excess, sugar can be toxic. Sufficient amounts of B-vitamins are actually required to metabolize and detoxify sugar in our bodies. When the body experiences a sugar overload, the assimilation of nutrients from other foods can be inhibited. In other words, our bodies were not designed to cope with the enormous quantity of sugar we routinely ingest. Eating too much sugar can generate a type of nutrient malnutrition, not to mention its contribution to obesity, diabetes, hyperactivity, and other disorders. Sugar can also predispose the body to yeast infections, aggravate some types of arthritis and asthma, cause tooth decay, and may even elevate our blood lipid levels.

Eating excess sugar can also contribute to amino acid depletion, which has been linked with depression and other mood disorders. To make matters worse, eating too much sugar can actually compromise our immune systems by lowering white blood cells counts. This makes us more susceptible to colds and other infections. Sugar consumption has also been linked to PMS, osteoporosis and coronary heart disease.

Why Do We Crave Sweets?

Considering the sobering effects of a high sugar diet, why do we eat so much of it? One reason is that sugar gives us a quick infusion of energy. It can also help to raise the level of certain brain neurotransmitters which may temporarily elevate our mood. Sugar cravings stem from a complex mix of physiological and psychological components. Even the most brilliant scientists fail to totally comprehend this intriguing chemical dependence which, for the most part, hurts our overall health.

What we do know is that when sugary foods are consumed, the pancreas must secrete insulin, a hormone which serves to bring blood glucose levels down. This allows sugar to enter our cells where it is either burned off or stored. The constant ups and downs of blood sugar levels can become exaggerated in some individuals and cause all kinds of health problems. Have you ever been around someone who is prone to sudden mood swings characterized by violent verbal attacks or irritability? This type of volatile behavior is typical of people who crave sugar, eat it and then experience sugar highs and lows. Erratic mood swings can be linked to dramatic drops in blood sugar levels.

Hypoglycemia: Sign of Hard Times?

It is rather disturbing to learn that statisticians estimate that almost 20 million Americans suffer from some type of faulty glucose tolerance. Hypoglycemia and diabetes are the two major forms of blood sugar disorders and can deservedly be called modern day plagues.

Hypoglycemia is an actual disorder that can cause of number of seemingly unrelated symptoms. More and more studies are pointing to physiological as well as psychological disorders linked to disturbed glucose utilization in brain cells. One study, in particular, showed that depressed people have overall lower glucose metabolism (Slagle, 22). Hypoglycemia occurs when too much insulin is secreted in order to compensate for high blood sugar levels resulting from eating sugary or high carbohydrate foods. To deal with the excess insulin, glucagon, cortisol and adrenalin pour into the system to help raise the blood sugar back to acceptable levels. This can inadvertently result in the secretion of more insulin and the vicious cycle repeats itself.

A hypoglycemic reaction can cause mood swings, fatigue, drowsiness, tremors, headaches, dizziness, panic attacks, indigestion, cold sweats, and fainting. When blood sugar drops too low, an overwhelming craving for carbohydrates results. To satisfy the craving and compensate for feelings of weakness and abnormal hunger, sugary foods are once again consumed in excess.

Unfortunately, great numbers of people suffer from hypoglycemic symptoms. Ironically, a simple switch from a high sugar diet to one that emphasizes protein can help. In addition, because sugar cravings are so hard to control, a product like stevia can be of enormous value in preventing roller coaster blood sugar levels. One Colorado internist states:

People who are chronically stressed and are on a roller coaster of blood sugar going up and down are especially prone to dips in energy at certain times of day. Their adrenals are not functioning optimally, and when they hit a real low point, they want sugar. It usually happens in mid-afternoon when the adrenal glands are at their lowest level of functioning. (Janiger, 71)

Our craving for sweets in not intrinsically a bad thing; however, what we reach for to satisfy that craving can dramatically determine how we feel. Stevia can help to satisfy the urge to eat something sweet without changing blood sugar levels in a perfectly natural way and without any of the risks associated with other non-nutritive sweeteners.

Diabetes: Pancreas Overload?

Diabetes is a disease typical of western cultures and is evidence of the influence that diet has on the human body. Perhaps more than any other disease, diabetes shuts down the mechanisms which permit proper carbohydrate/sugar metabolism. When the pancreas no longer secretes adequate amounts of insulin to metabolize sugar, that sugar continues to circulate in the bloodstream causing all kinds of health problems. The type of diabetes that comes in later years is almost always related to obesity and involves the inability of sugar to enter cells, even when insulin is present. Diabetes can cause blindness, atherosclerosis, kidney disease, the loss of nerve function, recurring infections, and the inability to heal.

Heredity plays a profound role in the incidence of diabetes, but a diet high in white sugar and empty carbohydrates unquestionably contributes to the onset of the disease. It is estimated that over five million Americans are currently undergoing medical treatment for diabetes and studies suggest that there are at least four million Americans with undetected forms of adult onset diabetes. Diabetes is the third cause of death in this country and reflects the devastating results of a diet low in fiber and high in simple carbohydrates.

Most of us start our children on diets filled with candy, pop, chips, cookies, doughnuts, sugary juice, etc. Studies have found that diabetes is a disease which usually plagues societies that eat highly refined foods. Because we live in a culture that worships sweets, the availability of a safe sweetener like stevia, which does not cause stress on the pancreas is extremely valuable. If sugar consumption was cut in half by using stevia to "stretch" sweetening power, our risk for developing blood sugar disorders like diabetes and hypoglycemia could dramatically decrease.

STEVIA: THE IDEAL SWEETENER?

For anyone who suffers from diabetes, hypoglycemia, high blood pressure, obesity or chronic yeast infections, stevia is the ideal sweetener. It has all the benefits of artificial sweeteners and none of the drawbacks. Stevia can be added to a variety of foods to make them sweet without adding calories or impacting the pancreas or adrenal glands. It can help to satisfy carbohydrate cravings without interfering with blood sugar levels or adding extra pounds.

Using stevia to create treats for children is also another excellent way to avoid weight gain, tooth decay and possible hyperactivity. While it may take some getting used to initially, stevia products are becoming easier to measure and better tasting.

Stevia's Unique Taste Sensation

When the whole leaf extract or powdered forms of stevia make contact with the tongue, the resulting taste can be described as a sweet flavor, with a slight licorice-like and transient bitter flavor. If stevia is used correctly with hot water or some other liquid, both those flavors will disappear. At this writing, researchers are working on a new extraction process that will preserve stevia's sweetening potency while minimizing any aftertaste associated with the herb.

Additional Therapeutic Benefits

Consider the following quote:

Stevia . . . is not only non-toxic, but has several traditional medicinal uses. The Indian tribes of South America have used it as a digestive aid, and have also applied it topically for years to heal wounds. Recent clinical studies have shown it can increase glucose tolerance and decrease blood sugar levels. Of the two sweeteners (aspartame and stevia), stevia wins hands down for safety. (Whitaker)

Stevia has a long history of medicinal use in Paraguay and Brazil and while many of the therapeutic applications of stevia are anecdotal, they must be considered in that they have spanned generations. Experts who work with indigenous cultures frequently find that traditional applications of folk medicine can be verified with scientific data.

Stevia and Blood Sugar Levels

Clinical tests combined with consumer results indicate that stevia can actually help to normalize blood sugar. For this reason, the herb and its extracts are recommended in some countries as an actual medicine for people suffering from diabetes or hypoglycemia. Recent studies have indicated that stevia can increase glucose tolerance while decreasing blood sugar levels.

Paraguayan natives have traditionally used stevia tea to regulate blood sugar. Stevia decoctions for diabetes are common and are usually prepared by boiling or steeping the leaves in water (Bonvie, 53). While scientific studies are certainly warranted, it is thought that disturbed blood sugar levels respond to stevia therapy while normal levels remain unaffected.

Stevia and Weight Loss

Stevia is an ideal dietary supplement for anyone who wants to lose or maintain their weight. Because it contains no calories, it can satisfy cravings for sweets without adding extra pounds. It is also thought that using stevia may decrease the desire to eat fatty foods as well. Appetite control is another factor affected by stevia supplementation. Some people have found that their hunger decreases if they take stevia drops 15 to 20 minutes before a meal. While scientific studies are lacking in this area, it is presumed that the glycosides in stevia help to reset the appestat mechanism found in the brain, thereby promoting a feeling of satiety or satisfaction.

Much of our nation's obesity epidemic is due to the over consumption of sugar-containing foods. Unfortunately, most sugary snacks are also loaded with fat, compounding the problem. When a sugar craving hits, anything will usually do. Doughnuts, candy bars, pies, pastries and cookies are considered high calorie, fattening foods. Using stevia to sweeten snacks and beverages can result making weight loss and management much easier.

High Blood Pressure

It is thought that taking stevia can result in lowering elevated blood pressure levels while not affecting normal levels. This particular application has not been researched, but its potential as a treatment for hypertension must be considered when assessing the value of herbal medicines for disease.

Microorganism Inhibitor?

Stevia is thought to be able to inhibit the growth of certain bacteria and other infectious organisms. Some people even claim that using stevia helps to prevent the onset of colds and flu. Tests have supported the antimicrobial properties of stevia against streptococcus mutans (Bonvie, 54). The fact that stevia has the ability to inhibit the growth of certain bacteria helps to explain its traditional use in treating wounds, sores and gum disease. It may also explain while the herb is advocated for anyone who is susceptible to yeast infections or reoccurring strep infections, two conditions that seem to be aggravated by white sugar consumption.

Oral Tonic

Stevia can be used as an oral tonic to prevent tooth decay and gingivitis. Stevia extracts are sometimes added to toothpaste or mouthwashes to initiate this effect. Stevia is used in some Brazilian dental products with the assumption that the herb can actually help to prevent tooth decay and retard plaque deposits (Bonvie, 53). Stevia offers the perfect sweetener for oral products like toothpastes and mouthwash, enabling them to be more palatable without any of the drawbacks of other sweeteners.

Digestive Aid

Brazilians have used stevia to boost and facilitate better digestion (Bonvie, 53). Again, while this therapeutic application remains unresearched, the fact that stevia has a long history of use as a gastrointestinal tonic must be acknowledged. Plant glycosides can exert numerous therapeutic actions in the human body.

Stevia and Skin Care

Whole leaf stevia or its by-products have been used to soften and tone the skin and to ease wrinkles and lines. Facial masks can be made by adding liquid to the powder, and liquid elixirs can be used as facial toners to help tighten the skin. Stevia concentrate in the form of drops has also been used directly on sores or blemishes to promote healing. For this reason, some advocates of stevia use it on other skin conditions such as eczema, dermatitis, or minor cuts or wounds. Stevia tea bags can be placed over the eyes to ease fatigue and to tone the skin. Stevia skin care products are available in clay bases, masks, and water-based creams. Liquid extracts can be directly applied to the skin.

FORMS OF STEVIA

Stevia has traditionally been used in either a powder or raw liquid form. Powdered forms can either be crude green or fine and white. Powders come in bulk or in tea bags. White stevia powder is the most common type and usually has more sweetening power than other forms. Countries like Japan use a filler substance along with stevia powder in order to give it more substance and make it easier to package. Powdered forms can be somewhat difficult to measure, although they are considered quite practical. Liquid formulas which are often brown in color frequently add other compounds to counteract bitterness. Alcohol based extracts are also available, as well as new concentrated liquid varieties. White stevia powder is the most popular form of the sweetener, although the leaf, ground or whole, can be purchased loose or in tea bags. Fresh leaves can be chewed but they are not practical for sweetening other foods. Dried leaves can used used for teas or in tea blends. Stevia tablets are also available for those who want to use the herb as a therapeutic rather than sweetening agent. Ground stevia can be sprinkled over cereals, salads, and other ready-to-serve foods. (NOTE: Stevia powders can vary in their sweetening strength, which is determined to a great degree by the refining process and the plant quality.)

If you choose to buy stevia leaves, they can widely vary in their quality and content depending on their cultivation and environmental conditions. The stevioside and rebaudidoside contents can also differ and bacterial or fungal contamination can be a problem. For this reason purchase stevia products only from reliable sources. Buying stevia in white powder or liquid extract forms from reliable distributors is also recommended.

Stevioside

Stevioside is the most powerful form of the stevia glycoside and is usually available in either a white powder or liquid extract. It is the isolated glycoside form of stevia and is used specifically for its sweetening ability and not for any therapeutic applications. Japanese consumers use stevioside extensively.

HOW TO USE STEVIA

The most frequent mistake people make with powdered stevia is measuring out too much. Very tiny amounts of the powder can greatly sweeten. Liquid extracts can be measured out in drops until the right amount of sweetening is achieved. Often just one half to one teaspoon of the liquid achieves the same effect as one cup of white sugar. If a powdered form is used, mixing it with hot water is recommended in order to create a more workable concentrate. Hot liquids seem to release the sweetening power of stevia more rapidly. This concentrate should be refrigerated and measured out with an eye dropper.

Baked goods sweetened with stevia do not brown as much, and using stevia in recipes with distinct flavors like lemon, cinnamon, carob, etc. achieves better results than adding it to blander food items. Baking with stevia takes some getting used to. Stevia can also be added to other sweeteners like honey to lower their caloric content. People who cook with stevia often add it to honey or molasses to potentiate sweetening power in smaller quantities.

Stevia works particularly well on dairy products, fruit dishes, beverages and fresh desserts. It can be combined with other sugars such as molasses, honey, maple syrup, fructose etc. in order to minimize their use. (NOTE: Stevia does not work well with yeast breads which require caloric forms of sugar to rise.

SAFETY

The FDA has not given stevia the "generally recognized as safe" label; however, the herbal compound has been used for hundreds of years without any recorded side effects. Japanese studies found that the sweetener consistently yielded a non-toxic status, even after extensive toxicity trials. The Japanese have used stevia for years with the approval of Japanese control agencies, and in Paraguay the herb has enjoyed hundreds of years of consumption with no reports of detrimental side effects. No anomalies have ever been observed in cell, enzyme, chromosomal or other significant physiological parameters during these toxicity tests. Stevia has not been associated with any form of cancer or birth defects. Stevia consumption in Japan was approximated at 170 metric tons in 1987 with no cases of documented side effects (Bonvie, 38).

Scientific Toxicology Studies

Comprehensive and tedious clinical studies in Japan have more than established the fact that stevia can be taken safely. One such study used over 450 rats who were fed stevia for up to two years with doses many times greater than human consumption. No changes were observed in organ weights, blood biochemistry, growth, appearance, or cellular function (Bonvie, 38). The Japanese have found no indication that stevia affects fertility or unborn children and have never linked it to cancer or other cellular mutations. (NOTE: Diabetics and people with other medical conditions should always consult their physician before using this or any other dietary supplement and should never alter or stop their medication unless advised to by their physician.)

PRIMARY APPLICATIONS OF STEVIA

- diabetes
- obesity
- plaque retardant
- hyperactivity
- high blood pressure
- carbohydrate cravings
- tobacco and alcohol cravings
- hypoglycemia
- indigestion
- dental health
- yeast infections
- oral health
- skin toning and healing

CONCLUSION

If you fall into the category of a consumer who is searching for an excellent natural sweetening agent which is safe, powerful, and calorie-free, stevia extracts should be first on your list. Ironically, while enormous quantities of aspartame and saccharine continue to be consumed in this country, a sweetening substance that poses less risk and is more effective continues to be rigorously regulated. Fortunately, restrictions are easing and it is now possible to purchase stevia as a supplement.

Both xylitol and saccharine have been linked to tumor development and aspartame continues to prompt controversy in its reported wide range of negative side effects, yet all of these products enjoy unrestricted marketability. It is rather ironic that chemical compounds that have the capability of wreaking all kinds of havoc with human physiology have the advantage over natural substances that are certainly much more benign. It's hard to imagine that a safe, natural herb which offers concentrated sweetening power and may also actually normalize blood sugar and prevent tooth decay remains relatively unknown.

Stevia will inevitably emerge as one of the best non-caloric sweeteners available. It's just a matter of time before American consumers discover its extraordinary attributes. In the meantime, learning to use stevia dietary supplements can provide us with the ability to "sweeten" our lives without compromising our health.

(NOTE: Linda Bonvie, Bill Bonvie and Donna Gates have written a comprehensive and engaging book on stevia called *The Stevia Story*. They have done extensive research and have put together a well written treatise on the subject. In addition, there are over fourteen current clinical studies on stevia listed in Medline which discuss various biochemical attributes of the herb's glycosides.)

BIBLIOGRAPHY

Aquino, R.P., "Isolation of the principal sugars of *Stevia rebaudiana*," *Boll Soc Ital Biol Sper ALS*,.Sept. 30, 1985, 61 (9): 1247-52.

Bertoni, Moises, "Kaa He-He, its nature and its properties," *Paraguayan Scientific Annals,* Dec. 10, 1905. See also Bonvie, 24.

Bonvie, Linda, Bill Bonvie and Donna Gates, *The Stevia Story,* (B.E.D., Atlanta, Georgia: 1997).

Gittleman, Ann, *Super Nutrition for Women,* (Bantam Books, New York: 1991).

Janiger, Oscar M.D., and Philip Goldberg, *A Different Kind of Healing,* (New York: Putnam and Sons, 1993).

Olney, John, *The Journal of Neuropathology and Experimental Neurology,* 1996.

Slagle, Patricia, M.D., *The Way Up From Down,* (New York: Random Books, 1987).

Smoliar, V. I., "Effect of a new sweetening agent from Stevia rebaudiana on animals," *Vopr Pitan XK4,* Jan-Feb., 1992, 60-63.

Weil, Andrew, *Natural Health, Natural Medicine,* (Houghton-Mifflin Co., Boston: 1990).

Whitaker Julian, MD, Newsletter, December. 1994.

Wurtman, R., "Aspartase effects on brain serotonin," *American Journal of Clinical Nutrition,* 1987, 45: 799-801.

Wurtman, R., "Neurochemical changes following high-dose aspartase with dietary carbohydrates," *New England Journal of Medicine,* 1983, 389: 429-30.

Wurtman, R., "Possible effect on seizure susceptibility," *Lancet,* 1985, 2: 1060.